COSMIC TREK

THE STORY OF EVERYTHING FROM BOUNDLESS UNIVERSE TO THE TINIEST PARTICLES

Written & Illustrated By
Sita Padmini

The Big Bang

Once upon a time, in a place so vast and mysterious, there was a great explosion. This explosion was so powerful that it created everything we see in the universe today. Scientists call this event the Big Bang and it happened about 13.8 billion years ago.

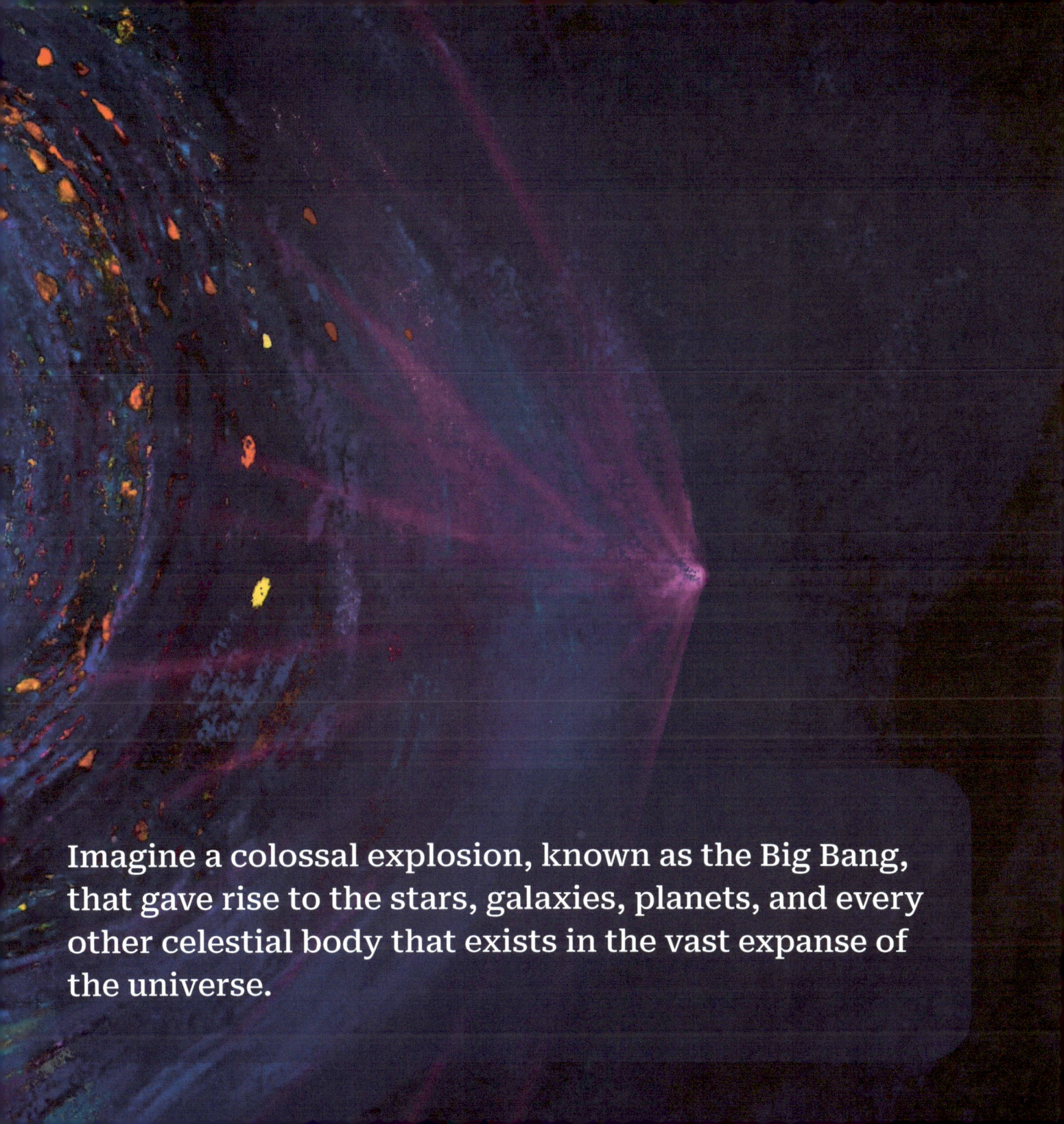

Imagine a colossal explosion, known as the Big Bang, that gave rise to the stars, galaxies, planets, and every other celestial body that exists in the vast expanse of the universe.

The Observable Universe

The universe is so huge that we can only see a tiny part of it. This part is called the Observable Universe and it contains stars, galaxies, and planets.

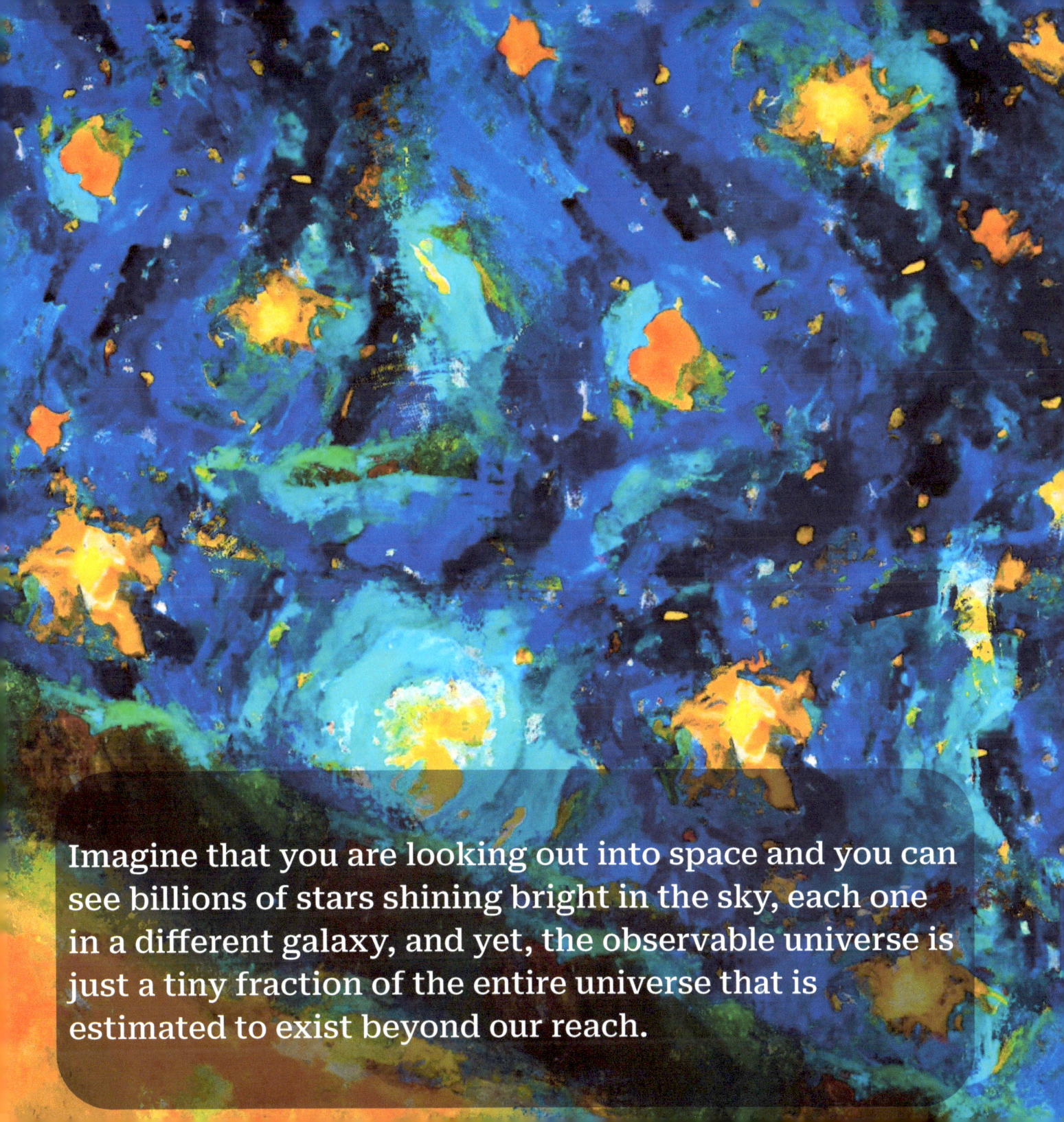

Imagine that you are looking out into space and you can see billions of stars shining bright in the sky, each one in a different galaxy, and yet, the observable universe is just a tiny fraction of the entire universe that is estimated to exist beyond our reach.

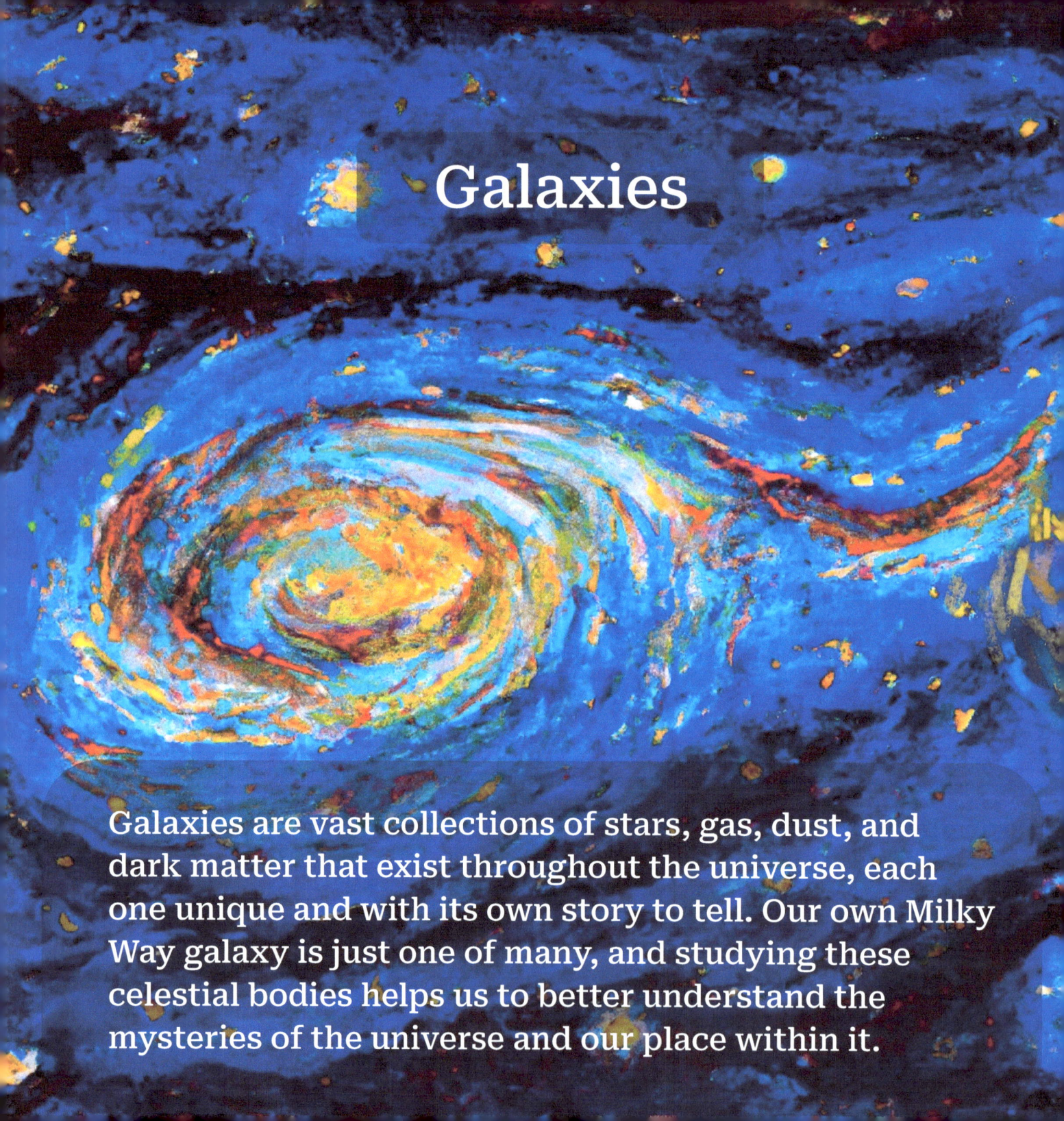

Galaxies

Galaxies are vast collections of stars, gas, dust, and dark matter that exist throughout the universe, each one unique and with its own story to tell. Our own Milky Way galaxy is just one of many, and studying these celestial bodies helps us to better understand the mysteries of the universe and our place within it.

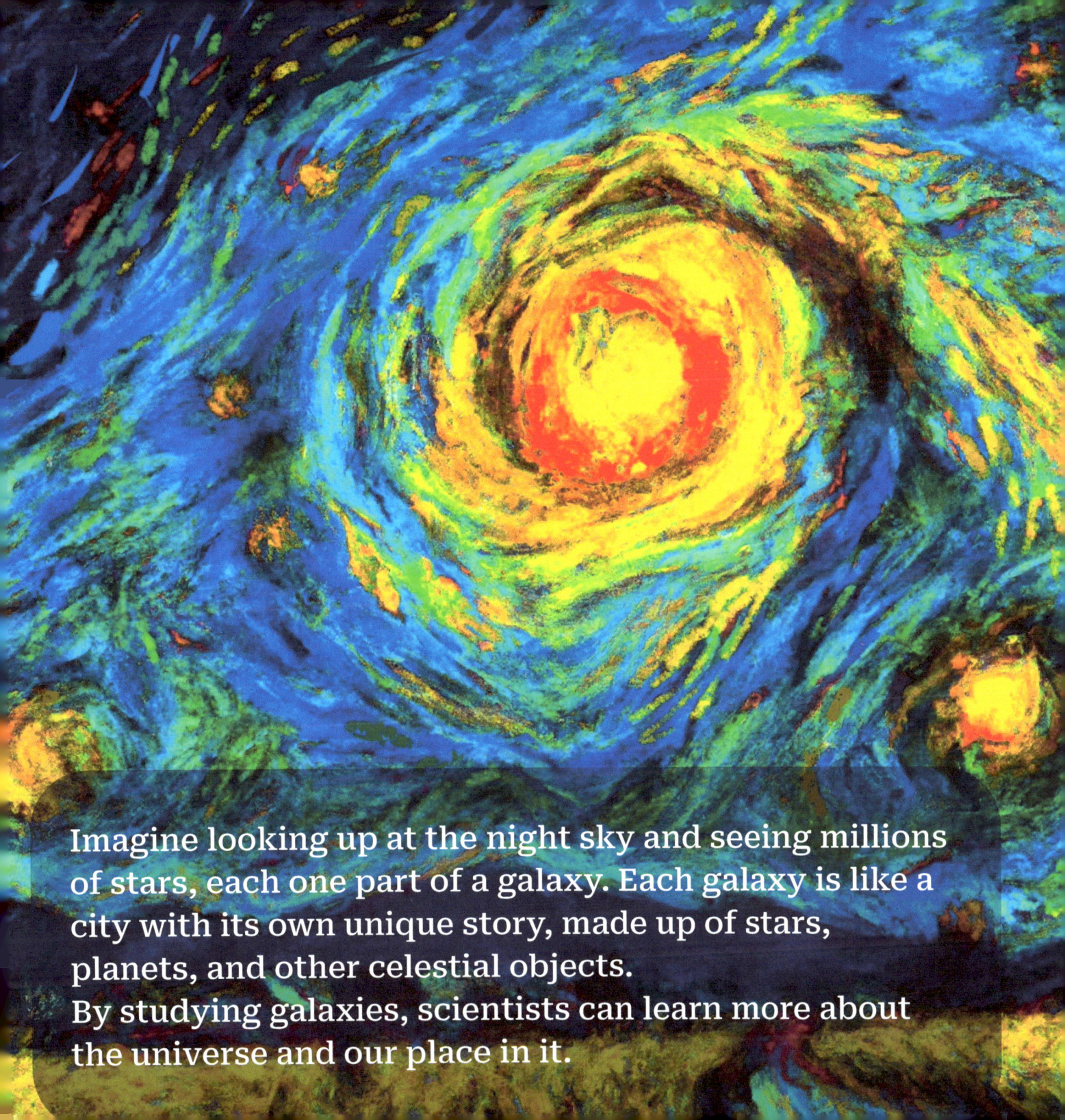

Imagine looking up at the night sky and seeing millions of stars, each one part of a galaxy. Each galaxy is like a city with its own unique story, made up of stars, planets, and other celestial objects.
By studying galaxies, scientists can learn more about the universe and our place in it.

Super Clusters

Super clusters are even larger collections of galaxies and galaxy clusters. They are some of the largest structures in the universe.

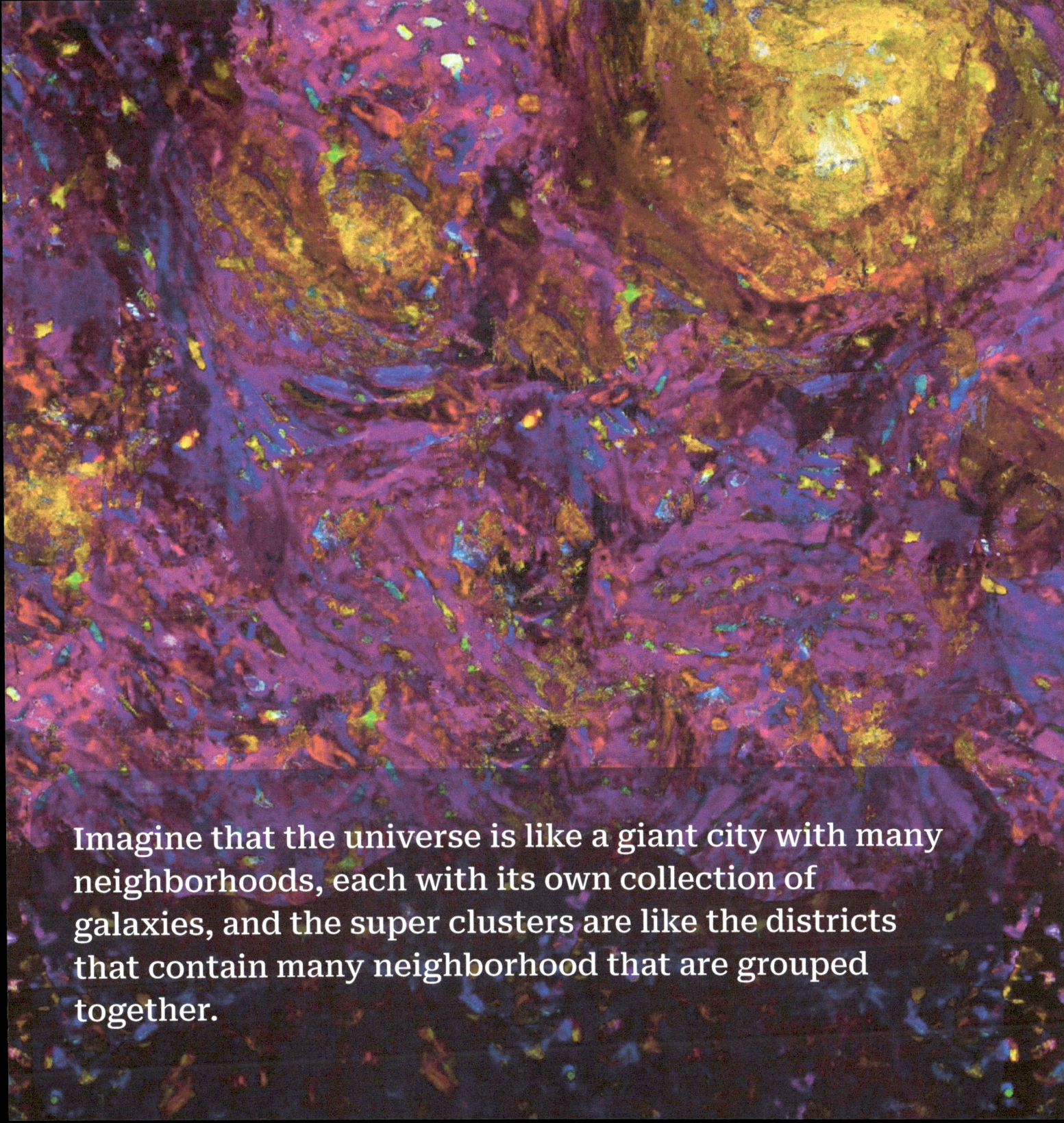

Imagine that the universe is like a giant city with many neighborhoods, each with its own collection of galaxies, and the super clusters are like the districts that contain many neighborhood that are grouped together.

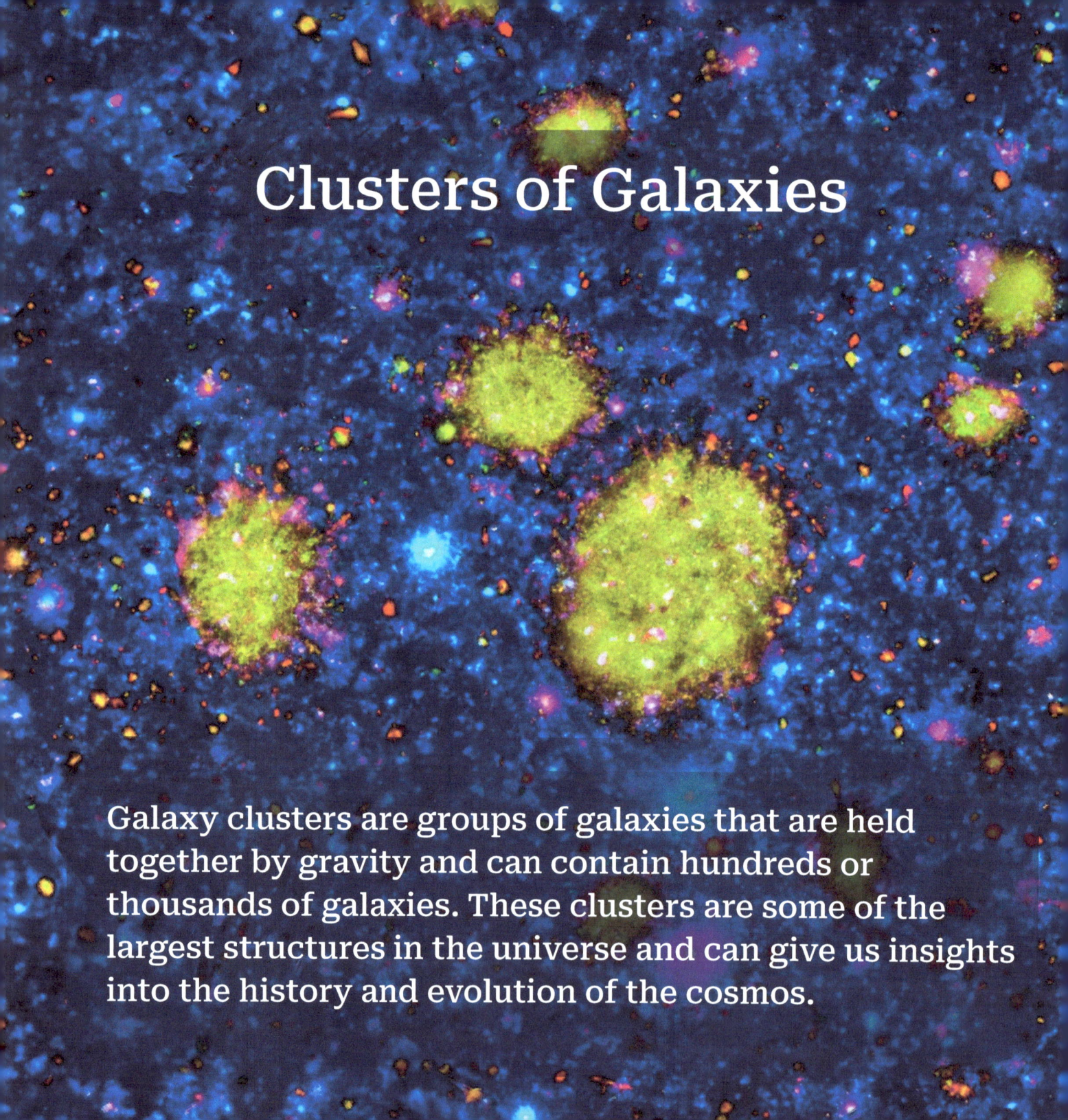

Clusters of Galaxies

Galaxy clusters are groups of galaxies that are held together by gravity and can contain hundreds or thousands of galaxies. These clusters are some of the largest structures in the universe and can give us insights into the history and evolution of the cosmos.

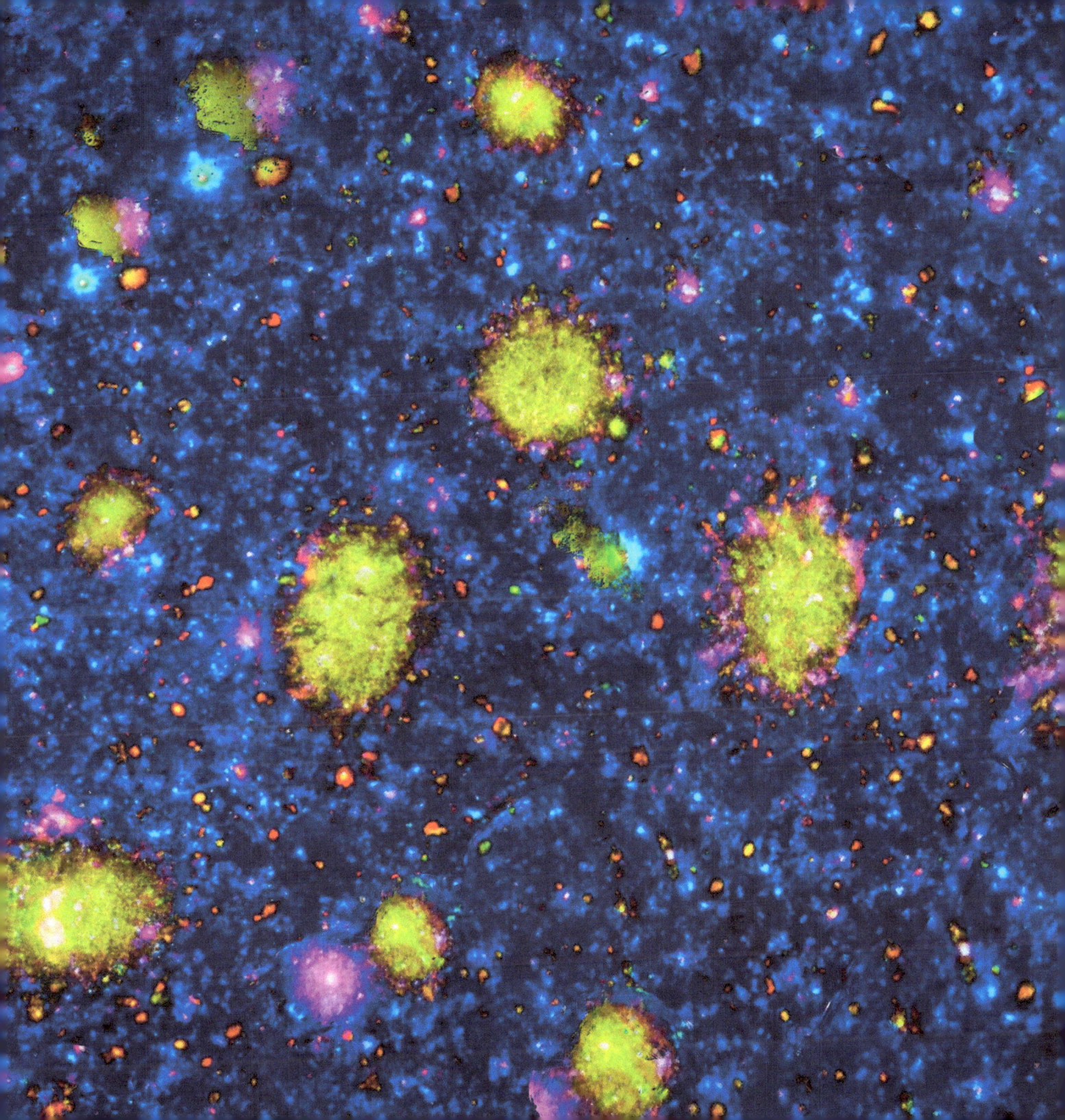

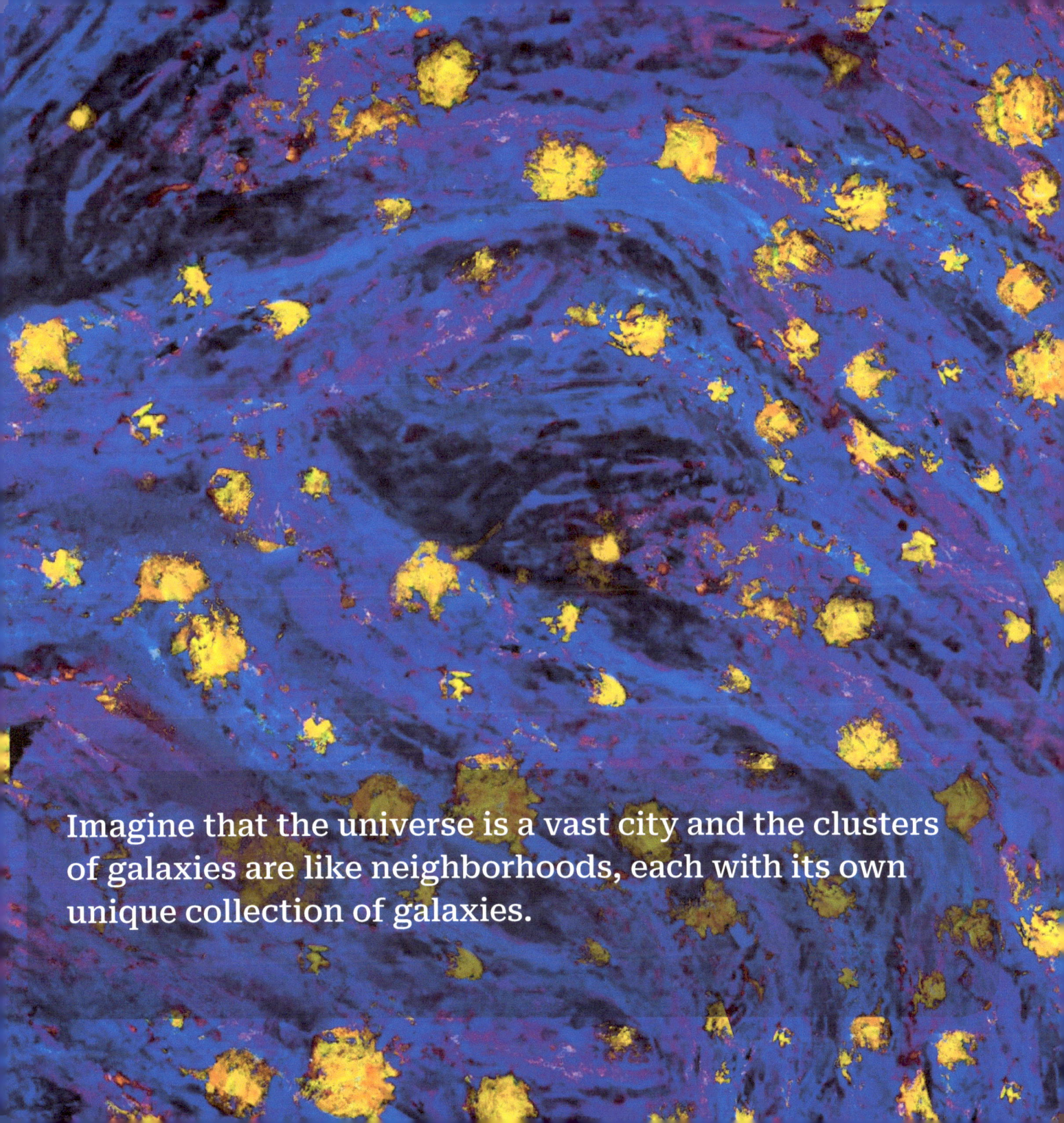

Imagine that the universe is a vast city and the clusters of galaxies are like neighborhoods, each with its own unique collection of galaxies.

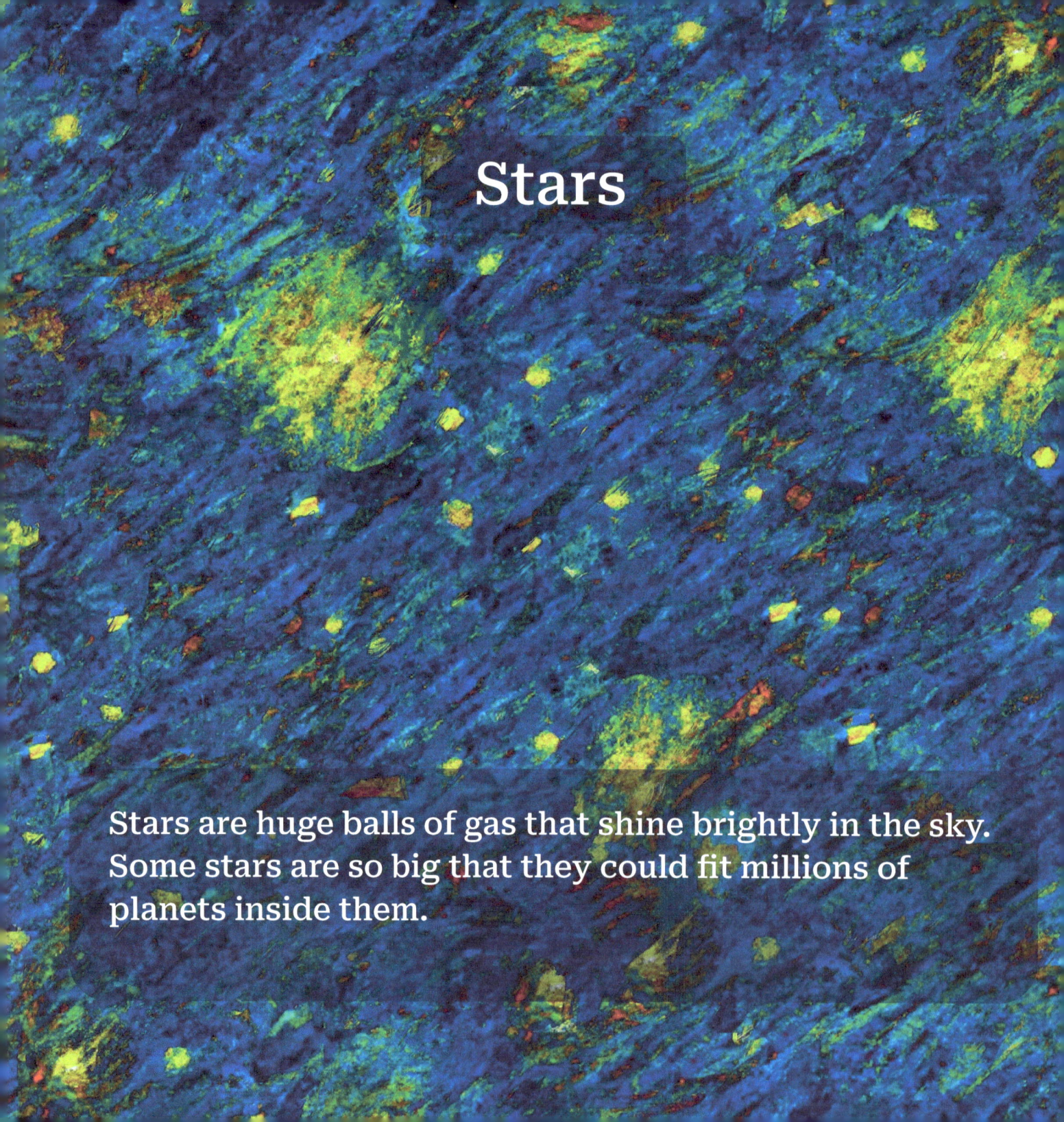

Stars

Stars are huge balls of gas that shine brightly in the sky. Some stars are so big that they could fit millions of planets inside them.

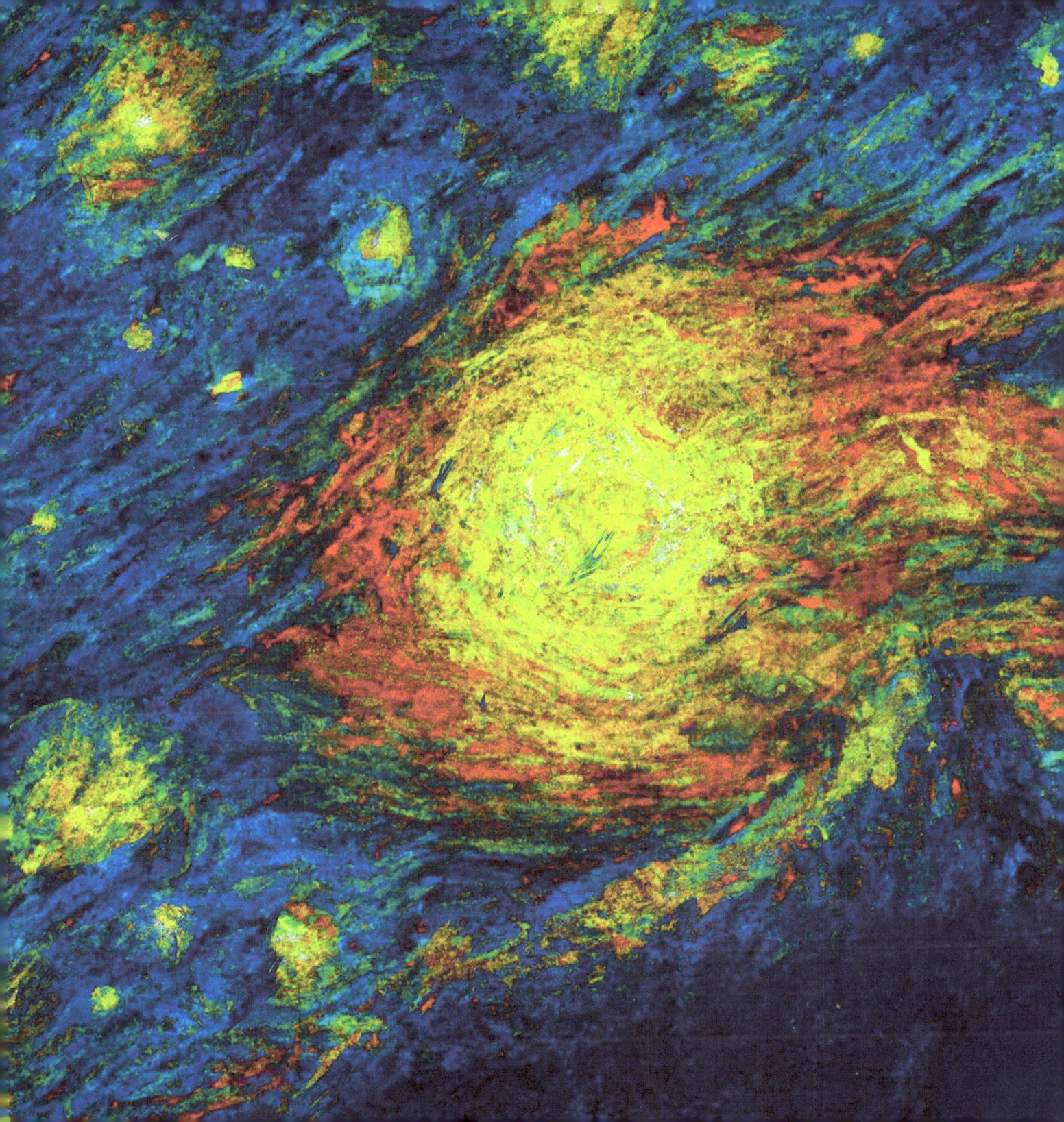

Imagine standing on a planet and looking up at the sky and seeing a bright star shining down on you. This star is so big that it could have its own system of planets.

Planets

Planets are celestial bodies that orbit around stars. Some planets are similar to Earth, while others are made of gas or ice.

Imagine embarking on a journey to explore various planets, where you encounter numerous breathtaking discoveries on each unique world.

Atoms

Atoms are the building blocks of everything in the universe. They are so small that you need a microscope to see them.

Imagine that everything you see and touch is made up of tiny particles called atoms. These atoms are what give things their properties and make up everything from the stars to the planets to you.

Protons, Neutrons, and Electrons

Atoms consist of even tinier particles known as protons, neutrons, and electrons, which are responsible for giving atoms their distinctive properties and make up everything that we perceive and interact with in the world around us.

Imagine that every single aspect of the world around us, from the tiniest grains of sand to the mightiest mountains, is formed by these minuscule particles working together in perfect harmony to create the reality we live in.

Quarks

Quarks are subatomic particles that combine to form protons and neutrons, the building blocks of atomic nuclei, and by extension, everything that we encounter in our everyday lives. They are so small that they can only be detected through experiments.

Imagine that the universe and everything in it, including us, are composed of these imperceptibly tiny particles called quarks, which are responsible for giving matter its unique properties and form.

The Higgs Boson

The Higgs boson is a subatomic particle that was discovered by scientists in 2012 using the Large Hadron Collider, a giant machine that smashes particles together at high speeds. The Higgs boson is important because it helps to explain how particles, such as protons and neutrons, get their mass. Without the Higgs boson, these particles would have no mass and the universe as we know it could not exist.

Imagine that the universe and everything in it, including us, are composed of these imperceptibly tiny particles called quarks, which are responsible for giving matter its unique properties and form.

Our journey through the cosmos is an eternal one, as scientists keep uncovering remarkable and unexplored aspects of our world. Every entity in the universe, starting from the Big Bang to the most minute particles, is interlinked and has a fascinating tale to narrate. Considering the boundless potential of discovery, let us continue our expedition to unearth the enigmatic wonders of the universe.

www.ingramcontent.com/pod-product-compliance
Lightning Source LLC
Chambersburg PA
CBHW040407220526
45473CB00004B/1160